About the Fraser Institute

The Fraser Institute is a social research and educational organization. It has as its objective the redirection of public attention to the role of competitive markets in providing for the well-being of Canadians. Where markets work, the Institute's interest lies in trying to discover prospects for improvement. Where markets do not work, its interest lies in finding the reasons. Where competitive markets have been replaced by government control, the interest of the Institute lies in documenting objectively the nature of the improvement or deterioration resulting from government intervention.

The Fraser Institute is a national, federally-chartered, non-profit organization financed by the sale of its publications and the tax-deductible contributions of its members, foundations, and other supporters; it receives no government funding.

Editorial Advisory Board

Prof. Armen Alchian
Prof. Jean-Pierre Centi
Prof. Michael Parkin
Prof. L.B. Smith

Prof. J.M. Buchanan
Prof. Herbert G. Grubel
Prof. Friedrich Schneider
Sir Alan Walters

Senior Fellows

Murray Allen, MD
Dr. Paul Brantingham
Prof. Barry Cooper
Prof. Herb Emery
Gordon Gibson
Prof. Ron Kneebone
Dr. Owen Lippert
Prof. Jean-Luc Migue
Dr. Filip Palda

Prof. Eugene Beaulieu
Martin Collacott
Prof. Steve Easton
Prof. Tom Flanagan
Dr. Herbert Grubel
Prof. Rainer Knopff
Prof. Ken McKenzie
Prof. Lydia Miljan
Prof. Chris Sarlo

Administration

Executive Director, Michael Walker
Director, Finance and Administration, Michael Hopkins
Director, Alberta Initiative, Barry Cooper
Director, Communications, Suzanne Walters
Director, Development, Sherry Stein
Director, Education Programs, Annabel Addington
Director, Publication Production, J. Kristin McCahon
Events Coordinator, Leah Costello
Coordinator, Student Programs, Vanessa Schneider

Research

Director, Fiscal and Non-Profit Studies, Jason Clemens
Director, School Performance Studies, Peter Cowley
Director, Pharmaceutical Policy Research, John R. Graham
Director, Centre for Studies in Risk and Regulation, Laura Jones
Director, Centre for Globalization Studies, Fred McMahon
Director, Education Policy, Claudia Rebanks Hepburn
Senior Research Economist, Joel Emes

Ordering publications

To order this book, any other publications, or a catalogue of the Institute's publications, please contact the book sales coordinator via our **toll-free order line: 1.800.665.3558, ext. 580**; via telephone: 604.688.0221, ext. 580; via fax: 604.688.8539; via e-mail: sales@fraserinstitute.ca.

Media

For media information, please contact Suzanne Walters, Director of Communications: via telephone: 604.714.4582 or, from Toronto, 416.363.6575, ext. 582; via e-mail: suzannew@fraserinstitute.ca

Website

To learn more about the Institute and to read our publications on line, please visit our web site at www.fraserinstitute.ca.

Membership

For information about membership: in **Vancouver**, please contact us via mail: The Development Department, The Fraser Institute, 4th Floor, 1770 Burrard Street, Vancouver, BC, V6J 3G7; via telephone: 604.688.0221 ext. 586; via fax: 604.688.8539; via e-mail: membership@fraserinstitute.ca

In **Calgary**, please contact us via telephone: 403.216.7175 or toll-free: 1.866.716.7175; via fax: 403.234.9010; via e-mail: barrym@fraserinstitute.ca

In **Toronto**, please contact us via telephone: 416.363.6575; via fax: 416.601.7322.

Publication

Editing and design by Kristin McCahon
and Lindsey Thomas Martin
Cover design by Brian Creswick @ GoggleBox.

Risk Controversy Series
General Editor, Laura Jones

The Fraser Institute's Risk Controversy Series publishes a number of short books explaining the science behind today's most pressing public-policy issues, such as global warming, genetic engineering, use of chemicals, and drug approvals. These issues have two common characteristics: they involve complex science and they are controversial, attracting the attention of activists and media. Good policy is based on sound science and sound economics. The purpose of the Risk Controversy Series is to promote good policy by providing Canadians with information from scientists about the complex science involved in many of today's important policy debates. The books in the series are full of valuable information and will provide the interested citizen with a basic understanding of the state of the science, including the many questions that remain unanswered.

An upcoming issue of the Risk Controversy Series will investigate misconceptions about the causes of cancer. Suggestions for other topics are welcome.

About the Centre for Studies in Risk and Regulation

The Fraser Institute's Centre for Studies in Risk and Regulation aims to educate Canadian citizens and policy-makers about the science and economics behind risk controversies. As incomes and living standards have increased, tolerance for the risks associated with everyday activities has decreased.

While this decreased tolerance for risk is not undesirable, it has made us susceptible to unsound science. Concern over smaller and smaller risks, both real and imagined, has led us to demand more regulation without taking account of the costs, including foregone opportunities to reduce more threatening risks. If the costs of policies intended to reduce risks are not accounted for, there is a danger that well-intentioned policies will actually reduce public well-being. To promote more rational decision-making, the Centre for Studies in Risk and Regulation will focus on sound science and consider the costs as well as the benefits of policies intended to protect Canadians.

For more information about the Centre, contact Laura Jones, Director, Centre for Studies in Risk and Regulation, The Fraser Institute, Fourth Floor, 1770 Burrard Street, Vancouver, BC, V6J 3G7; via telephone: 604.714.4547; via fax: 604.688.8539; via e-mail: lauraj@fraserinstitute.ca

Risk Controversy Series 2

Biotechnology & Food for Canadians

Alan McHughen

The Fraser Institute
Centre for Studies in Risk and Regulation
Vancouver   British Columbia   Canada   2002

American Council on Science and Health
New York City   United States of America   2002

Copyright ©2002 by The Fraser Institute. All rights reserved. No part of this book may be reproduced in any manner whatsoever without written permission except in the case of brief passages quoted in critical articles and reviews.

This publication is based on *Biotechnology and Food* (second edition), published in September 2000 by the American Council on Science and Health (2nd floor, 1995 Broadway, New York NY 10023 USA; www.acsh.org). It was updated and adapted for Canada by the author.

The author of this book has worked independently and opinions expressed by him are, therefore, his own and do not necessarily reflect the opinions of the members or the trustees of The Fraser Institute.

Printed in Canada.

National Library of Canada Cataloguing in Publication Data

McHughen, Alan
Biotechnology & food for Canadians

    (Risk controversy series ; 2)
    Copublished by: American Council on Science and Health
    Includes bibliographical references.
    ISBN 0-88975-191-9

    1. Genetically modified foods—Canada. 2. Food—Biotechnology—Canada. I. Centre for Studies in Risk and Regulation. II. American council on Science and Health. III. Title. IV. Series.
TP248.65.F66M33 2002   363.19'29'097   C2002-910018-6

Contents

About the author   /   ix

Foreword   /   xi

Introduction   /   3

Genetic modification of food   /   13

Concerns about GM foods and products   /   29

Conclusion   /   49

Appendix   /   51

Glossary   /   59

References and further reading   /   65

About the author

Alan McHughen is a public-sector educator, scientist and consumer advocate. After earning his doctorate at Oxford University and working at Yale University, Dr McHughen spent 20 years as Professor and Senior Research Scientist at the University of Saskatchewan before joining the faculty at the University of California. A molecular geneticist with an interest in crop improvement, he has helped develop Canada's regulation covering the environmental release of plants with novel traits.

He also served on a recent OECD panels investigating the effects upon health of genetically modified foods. Having developed internationally approved commercial crop varieties using both conventional breeding and genetic engineering techniques, he has first-hand experience with all the relevant issues from both sides of the regulatory process, covering both recombinant DNA and conventional breeding technologies. He also serves on the national expert committee on variety registration (Including several years on the executive panel as Secretary of the Oilseeds Subcommittee). He is currently acting Chair of the International Society for Biosafety Research and serves on a panel of the US National Academy of Sciences that is reviewing the American regulatory framework for genetically engineered plants.

As an educator and consumer advocate, he helps non-scientists understand the environmental and health impacts of both modern and traditional methods of food production. His award winning book, *Pandora's Picnic Basket: The Potential and Hazards of Genetically Modified Foods* (Oxford University Press, ISBN 0-19-850674-0), uses understandable, consumer-friendly language to explode the myths and explore the genuine risks of genetic modification (GM) technology.

Foreword

Biotechnology & Food for Canadians is the second publication in The Centre for Studies in Risk and Regulation's Risk Controversy Series, which will explain the science behind many of today's most pressing public-policy issues. Many current public-policy issues such as global warming, genetic engineering, use of chemicals, and drug approvals have two common characteristics: they involve complex science and they are controversial, attracting the attention of activists and media. The mix of complex science, activists' hype, and short media clips can bewilder the concerned citizen.

The activists

The development and use of new technology has long attracted an "anti" movement. Recent high-profile campaigns include those against globalization, genetic engineering, cell phones, breast implants, greenhouse gases, and plastic softeners used in children's toys. To convince people that the risks from these products or technologies warrant attention, activists rely on dramatic pictures, public protests, and slogans to attract media attention and capture the public's imagination. The goal of these campaigns is not to educate people so they can make informed choices for themselves—the goal is to regulate or, preferably, to eliminate the offending product or technology. While activists' personal motivations vary, their campaigns have three common characteristics. First, there is an underlying suspicion of economic development. Many prominent environmental activists, for example, say that economic growth is the enemy of the environment and among anti-globalization crusaders, "multinational corporation" is a dirty word. Second, the benefits of the products, technologies, or life-styles that activists attack are ignored while the risks are emphasized and often exaggerated. Many activists insist that a product or technology be proven to pose no risk at all before it is brought to market—this is sometimes called the precautionary principle. This may sound sensible but it is, in fact, an absurd demand: nothing, including many products that we use and activities we enjoy daily, is completely safe. Even the simple act of eating an apple poses some risk—one could choke on the apple or the apple might damage a tooth. Finally, activists have a tendency to focus only on arguments that support their claims, which often means dismissing legitimate scientific debates and ignoring uncertainty: activists claim, for example, that there is a consensus among scientists that global warming is caused largely by human activity and that something must therefore be done to control greenhouse gas emissions. As the first publication in this series showed, no such consensus exists.

The media

Many of us rely exclusively on the media for information on topics of current interest as, understandably, we do not have time to conduct our own, more thorough literature reviews and investigations. For business and political news as well as for human-interest stories, newspaper, radio, and television media do a good job of keeping us informed. But, these topics are relatively straight-forward to cover as they involve familiar people, terms, and places. Stories involving complex science are harder to do. Journalists covering these stories often do not have a scientific background and, even with a scientific background, it is difficult to condense and simplify scientific issues for viewers or readers. Finally, journalists work on tight deadlines, often having less than a day to research and write a story. Tight deadlines also make it tempting to rely on activists who are eager to provide information and colourful quotations.

Relying on media for information about a complex scientific issue can also give one an unbalanced view of the question because bad news is a better story than good news. In his book, *A Moment on the Earth*, Gregg Easterbrook, a reporter who has covered environmental issues for *Newsweek*, *The New Republic*, and *The New York Times Magazine*, explains the asymmetry in the way the media cover environmental stories.

> In the autumn of 1992, I was struck by this headline in the *New York Times*: "Air Found Cleaner in US Cities." The accompanying story said that in the past five years air quality had improved sufficiently that nearly half the cities once violating federal smog standards no longer did so. I was also struck by how the *Times* treated the article—as a small box buried on page A24. I checked the nation's other important news organizations and learned that none had given the finding prominence. Surely any news that air quality was in decline would have received front-page attention (p. xiii).

Despite dramatic overall improvements in air quality in Canada over the past 30 years, stories about air quality in Canada also focus on the bad news. Both the *Globe and Mail* and the *National Post* emphasized reports that air quality was deteriorating. Eighty-nine percent of the *Globe and Mail*'s coverage of air quality and 81 percent of the *National Post*'s stories in 2000 focused on poor air quality (Miljan, Air Quality Improving—But You'd Never Know It from the *Globe & Post*, *Fraser Forum*, April 2001: 17–18).

That bad news makes a better story than good news is a more generally observable phenomenon. According to the Pew Research Center for the People and the Press, each of the top 10 stories of public interest in the United States during 1999 were about bad news. With the exception of the outcome of the American election, the birth of septuplets in Iowa, and the summer Olympics, the same is true for the top 10 stories in each year from 1996 through 1998 (Pew Research Center for the People and the Press 2000, digital document: www.people-press.org/yearendrpt.htm).

While it is tempting to blame the media for over-simplifying complicated scientific ideas and presenting only the bad news, we must remember that they are catering to the desires of their readers and viewers. Most of us rely on newspapers, radio, and television because we want simple, interesting stories. We also find bad news more interesting than good news. Who would buy a paper that had "Millions of Airplanes land safely in Canada each Year" as its headline? But, many of us are drawn to headlines that promise a story giving gory details of a plane crash.

The Risk Controversy Series
Good policy is based on sound science and sound economics. The purpose of the Risk Controversy Series is to promote good policy by providing Canadians with information from scientists about the complex science involved in many of today's important policy debates. While these reports are not as short or as easy to read as a news story, they are full of valuable information and will provide the interested citizen with a basic understanding of the state of the science, including the many questions that remain unanswered.

An upcoming issues of the Risk Controversy Series will investigate misconceptions about the causes of cancer. Suggestions for other topics are welcome.

Laura Jones, Director
Environment and Regulatory Studies
Centre for Studies in Risk and Regulation

Biotechnology & Food
for Canadians

Introduction

Biotechnology and the consumer

Biotechnology is simply using living systems to give society more or better foods, drugs and other products. In this sense, we have been applying biotechnology since the dawn of civilization. In recent times, our understanding of science, especially genetics, has advanced to the point where we can optimize specific genes and traits to provide even greater benefits while reducing or eliminating undesirable features. Biotechnology, based on recombinant DNA (rDNA), is often called Gene Splicing, Genetic Engineering (GE) or Genetic Modification (GM), giving rise to a genetically modified organism (GMO).* The precision attained by such molecular plant breeding can provide, for example, greatly increased crop production and nutritional enhancements at little or no additional cost. Fruits and vegetables can be picked and delivered at the height of flavor and ripeness thanks to carefully tailored improvements that reduce softening and bruising. For health-conscious consumers, cooking oils from GM

---

* While the term "gene splicing" is more technically correct, most consumers are familiar with the terms "genetic modification" or "GM" to signify changing genomes or organisms by inserting or deleting genes. Thus, in this publication, we will use the more familiar terminology of genetic modification.

corn, soy or canola will provide lower saturated fat content. Any interest in French fries with fewer calories? GM potatoes with enhanced starch content absorb less fat during frying. Leaner meats will be available from cattle and pigs improved both directly and through improved feeds. Sensitive new testing kits can detect tiny amounts of potentially harmful toxic contaminants in foods. New plant varieties that are biologically protected against insects and diseases are now on the market, just in time to help farmers hard pressed to maintain efficient production with fewer chemical control agents. As our knowledge of molecular genetics increases, our ability to improve our foods and farming will increasingly benefit consumers at home and around the world. Among the benefits to consumers is more nutritious food, more diverse foods, less expensive food and, in regions of most crucial need, more abundant food.

Unfortunately, there is much misinformation, misunderstanding and confusion about this technology. These circumstances give rise to needless anxiety and, at the same time, obscure any real hazards that might exist as well as possible means of controlling them. A basic understanding of the techniques and goals of biotechnology research is important for deciding the merits of concerns and proposed solutions. *Biotechnology & Food for Canadians* provides an overview of what is now available through modern biotechnology, what is in the pipeline, what is on the drawing board and how products of biotechnology are regulated by various government agencies. This publication is not intended to cover all issues and concerns in depth but to discuss briefly various salient points. References to sources for further reading are presented as a guide for those wishing to delve deeper into particular areas.

Real informed choice requires real information. This booklet explains the facts behind GM and explores some of the issues surrounding the increasingly contentious issues. Armed with facts, we can identify and discuss the actual benefits as well as methods for managing or avoiding any potential risks.

Biotechnology—the background facts

Biotechnology has given us almost all of our foods, from corn and cattle (via various traditional plant and animal breeding technologies) to bread and wine (from traditional fermentation technologies). The era of modern biotechnology started in the early 1970s, when American scientists Herb Boyer and Stan Cohen developed recombinant DNA (rDNA), or "gene splicing" methods, in which fragments of DNA are joined together to create a new genetic combination.

DNA is the genetic molecule inside the cells of bacteria, plants and animals, including humans. DNA carries the genes, which hold genetic information, much like a recipe. A gene is a unit of genetic information; it tells the cell how to make a specific protein. It is the presence or absence of the specific protein that gives an organism a trait or characteristic. Many common genes are already shared by many different species and rDNA allows us to transfer genes from one organism to another, even across the usual barriers between species faced by conventional breeders.

Over the past quarter-century, genetic modification, or rDNA technology, has given us lifesaving drugs like Humulin™ (human insulin); Pulmozyme™ (dornase alpha), a breakthrough treatment for cystic fibrosis; Betaseron™ (interferon beta-1b), a powerful new drug for certain multiple sclerosis patients; and Activase™, a clot-dissolving tissue plasminogen activator, a treatment for heart attacks. It has also provided a range of precise genetic diagnostic tools to identify, at an early stage, muscular dystrophy and AIDS, among other conditions. Genetic technology has also given us safer medical treatments. For example, the standard hepatitis-B vaccine was derived from blood pooled from people who had had hepatitis B. The problem was that before AIDS was recognized as a blood-borne disease, some donors might also have had HIV. As hepatitis-B vaccines from modern rDNA technology are not derived from human blood, unknown blood infections cannot contaminate the rDNA vaccines. The application of rDNA to medical problems was rapidly embraced by researchers and by the public. However, the same technology, applied to agriculture, is facing resistance by some people who think it might be inherently hazardous. However, since the basic technology is the same, it is difficult to see why it might be hazardous to use biotechnology to make foods but not to make medicines. In any case, the distinction is becoming blurred: biotechnology is used, for example, to develop plants that make medicines. A leading scientist in this is Calgary's Dr. Maurice Moloney, whose team is busy genetically modifying crop plants to make drugs.

The first GM plants were produced in 1983 and food scientists lost no time in applying GM technology to improving crops. Biotechnology provides new tools for scientists working on long-standing agricultural problems in pest and disease management, animal and crop yield, and food quality. These tools complement and extend traditional selective-breeding techniques by providing the means for making selective, single-gene changes in plants and animals. In contrast, offspring created through conventional breeding present a random combination of thousands of genes from each parent. The differences between the traditional and modern methods are precision, speed and certainty. Moreover, because DNA is biochemically equivalent in all organisms, the modern techniques also enable scientists to take advantage of the full spectrum of genes present in nature—genes derived from microbes, plants and animals—in their efforts to improve agriculture. So, while the goals of traditional breeding and modern genetic engineering are similar, the new techniques greatly expand the range of possible strategies by eliminating the interspecies barriers presented by sexual reproduction. Adding one or two specific genes to a crop plant can help it fight off viral, bacterial or fungal diseases. Add other genes and the plant can produce biological insecticides to ward off pests. Still other genes will make the plant resistant to herbicides commonly used in weed control. Other strategies, with the puzzling names "antisense DNA," and "co-suppression" use GM to shut down production of a plant's own molecules. "Antisense DNA" produced the first GM whole food on the market, the *Flavr-savr*™ tomato, by inverting a natural gene for a ripening enzyme, resulting in tomatoes with extended shelf-life. In this gene-inversion process, the relevant gene is cut out of the DNA, turned 180 degrees, and reinserted; as if a phrase from this sentence were placed in reverse orientation, like this: "AND eht fo tuo tuc si eneg tnaveler eht."

Genetic engineering also provides sensitive diagnostics for veterinary medicine, *e.g.* for disease detection and gender identification. New animal vaccines are being developed, including vaccines with activity against more diverse strains of pathogens for diseases such as rabies that pose serious public-health threats, and those used to protect wild animals from diseases like the devastating rinderpest. An exciting collaboration involving the University of British Columbia, the University of Saskatchewan and the Alberta Research Council is developing vaccines against the devastating bacterial pathogen *E. coli* 0157:H7. Animals are being bred with additional genes for growth hormones—hormones that increase the ratio of protein to fat in the meat of pigs and cattle or the rate of milk production by dairy cows. Modern biotechnology is well established in food processing—particularly in the genetic improvement of bacteria and yeast strains used in various fermentation systems including improved bread yeast and brewing yeast. Vegetarians might now enjoy cheese made with GM chymosin instead of rennin from animals. Some recent developments in agricultural biotechnology involve using plants and animals to make products not traditionally associated with agriculture—products such as medically important pharmaceuticals and industrial materials to replace petroleum-based oils and plastics. In the future, we will all benefit from enhancements to our quality of life, environmental stewardship and animal welfare.

Canadian farmers have not had an easy time in recent years, with low prices for traditional commodities like wheat and barley combined with unpredictable and uncontrollable environmental events such as drought. Biotechnology, however, may open entirely new markets for their products and provide new products for their current markets, opening the opportunity for farming, and farmers, to become more profitable. One of the few bright spots for Canadian farmers over the past 30 years has been the development of canola, the high-quality vegetable oil derived from the rapeseed plant by a team of Canadian scientists led by Dr Keith Downey and Dr Baldur Stefansson. Although canola was developed using conventional plant-breeding methods, it showed what innovative technology and goal-oriented research could achieve. Canola is now Canada's second largest crop and the reason many farmers are still in business. Canola is also the target of much of Canada's biotechnology effort. According to the Canola Council of Canada, over 80% of canola farmers are growing genetically modified varieties; in so doing, they save millions of litres of fuel and tonnes of herbicide and increase their productivity and profitability. At the same time, the reduced fuel and herbicide use is better for the environment, so all citizens benefit.

Modern biotechnology has sparked both optimism and controversy. Debate about the impact of the research tools and products of biotechnology encompasses health, farm economics, global biodiversity, environmental quality and, not unimportantly, hunger and malnutrition.

DNA, genes and proteins

Deoxyribonucleic acid (DNA) is the hereditary "molecule of life" that carries the recipes for creating an organism. DNA, a threadlike molecule composed of long stretches of the small chemical compounds—usually called "bases"— Adenine, Thymine, Cytosine and Guanine (often abbreviated A, T, C and G, respectively), arranged in a particular order (see figure 1). Although the DNA of all organisms uses these same bases, the sequence of bases accounts for differences among species and even among individuals of the same species. We can think of the bases as letters in a language: where English uses 26 letters, DNA uses only four. Like English, DNA makes different words by placing the letters in different order. Differences in base sequence account for all the genetic differences among all living things. The DNA base sequence of apparently unrelated humans is remarkably similar. The base sequence of the DNA of identical twins is also identical while the DNA of fraternal twins differs as much as it does between other siblings.

Genes are portions of DNA providing the organism with information on how to make a specific protein. It is the presence or absence of a specific protein that gives an organism a particular trait. As DNA is passed from one generation to the next, so are the specific genes, allowing us to "inherit" traits from our parents. Throughout the history of life on earth, genes have carried information from one generation to the next. Genes are the fundamental drivers of evolution. Over millions of years, the evolution of complex organisms—plants, animals and humans—was achieved through the transfer, deletion and mutation of genes. Without genetic variability, evolution could not occur and the world would be devoid of most, if not all, life. In a sense, every organism on earth is the product of genetic engineering by nature. Each of us is in fact the product of an experiment in genetic engineering performed by our ancestors. The same is true for other animals, plants and microbes.

These natural mechanisms for genetic change allow an organism to gain new genes, new traits, and drive evolution. Biotechnology, both traditional and modern, simply takes advantage of these natural genetic phenomena to produce useful organisms and products from those organisms.

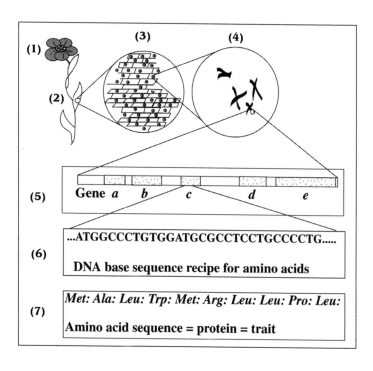

Figure 1 This diagram shows, using progressive magnification from a typical organism **(1)**, how the genes in every living cell carry instructions for specific traits. A plant leaf **(2)** is magnified to show the cells, each containing a round nucleus **(3)**. A magnifiction of the nucleus shows the chromosomes **(4)**. All higher organisms are made of cells, and all living cells carry nucleii. Chromosomes, which are composed of DNA and proteins, are carried inside every nucleus. If a portion of the DNA from a chromosome is stretched out, we can visualize how the genes are arranged **(5)**. Here, five genes (*a, b, c, d,* and *e*) are shown; each is a recipe for a different protein, and each is a different segment and length of DNA, with some interspersed DNA not being part of any gene. A further magnification of gene *c* **(6)** shows a portion of the DNA base sequence. The DNA sequence of each gene will differ but all use the same four base "letters." The cell machinery reads the recipe three letters at a time to make a protein. Panel **(7)** shows the amino-acid sequence as dictated by the DNA base sequence. ATG calls for the amino acid Methionine followed by Alanine from the next three bases, GCC. Each succeeding triplet of bases calls for a specific amino acid; our example continues with Leucine, Tryptophan, then—not shown in **(7)**—another Methionine, Arginine, Leucine, Leucine, Proline and Leucine. The cell machinery finds the appropriate amino acid in the cell and attaches it to the preceding one, creating a long chain, the basic structure of the protein. An average gene consists of about 1000 bases, translatign to over 300 amino acids in the resulting protein. The example here is the start of the actual DNA base sequence and amino-acid sequence of human insulin. (Adapted, with permission from Oxford University Press, from McHughen 2000.)

Genetic modification of food

Yesterday's biotechnology
Gene modification is a natural event. Almost all of our traditional foods are products of natural mutations or genetic recombinations. Even ordinary wheat for making bread, for example, is a product of the combination of DNA from three different species. This genetic modification, directed by Nature over the course of eons, gave us a more productive, palatable and nutritious food than was provided by any of the three originating species alone. Today, wheat is Canada's largest crop but it does not grow here naturally; it was introduced here by the pioneers. Similarly, modern corn looks almost nothing like teosinte (figure 2), its primitive genetic forebear. Conventional breeders combine genes from plants around the world, resulting in plants that would never exist under "natural" conditions. Purists who insist on eating only "natural" foods would have to avoid eating every one of our major agricultural products. Even the human genome (the complete collection of genetic information) carries on it remnants of viral genes deposited by the passing pathogens in our ancestors thousands of years ago. Natural gene transfer is not limited to evolutionary time—it is an on-going process and occurs daily. Agrobacterium is a soil-borne microbe with the natural ability to transfer pieces of

(1)   (2)   (3)   (4)

Figure 2   Traditional plant biotechnology has had profound impacts on our food crops. This photograph shows **(1)** the ancestor of modern corn, teosinte and **(4)** an ear of what we would recognize as corn today. Images **(2)** and **(3)** show hybrids, the result of crossing teosinte and modern corn. It took many generations to derive modern corn from the ancestral teosinte. (Photograph courtesy of Dr. John Doebley, University of Wisconsin.)

DNA into plant cells, where the bacterial genes are inserted into the plant genome; it does so every day without any human intervention. Humans occasionally, if inadvertently, eat these naturally occurring GM plants and have done so ever since humans started eating plants. Other forms of natural genetic modification result in changes to the genome similar to human-mediated modification. A transposon is a naturally occurring piece of DNA with the ability to excise itself from one part of the genome and insert itself somewhere else. Spontaneous mutations are also natural phenomena resulting in sometimes dramatic changes in the genetic information of an organism.

Conventional breeding practices

Genetic modification directed by humans started about 10,000 years ago, when human agricultural activities began. Traditional methods of modifying food products have been very successful, introducing substantial improvements to generate crops that would not otherwise occur in nature. Humans developed methods of introducing genetic material into crops to provide us with the remarkable diversity of fruits, vegetables and animals we enjoy today. They also provide the incredible volume of food required to allow the world population to grow to its present six billion. Traditional selective breeding has given us crops and strains with combinations of traits that would not occur without human intervention—pumpkins, potatoes, sugar beets, corn, oats, rice and black currants are common examples. Using cross-breeding methods, breeders transfer the complete set of genes from the parent plants to the new offspring. This introduces not only the one or two desired useful genes but thousands of other genes also, some benign and some undesirable. Breeders have developed effective but often expensive and time-consuming methods to eliminate the unwanted genes and to retain the desired improvements. Some conventional breeding programs involve quite dramatic gene shuffling: for example, most of our modern corn varieties are "double-cross hybrids" created when variety *A* is crossed with *B* to produce *C*, variety *D* is crossed with *E* to make *F*, which is then crossed with *C* to produce the desired seed.

Other "traditional" methods of plant breeding include mutation breeding, using ionizing radiation or other mutagenic agents to create genetic changes. Over 2,000 crop varieties around the world have been developed using mutation breeding, with its ability to alter DNA in unpredictable ways. In the 50 years that people have been eating these induced mutant foods, there are no recorded incidents of harm, even though the exact changes to the DNA remains unknown.

As useful as conventional genetic techniques have proved to be, they are also limited in their capability. If a desired gene is not available in a particular species or cannot be mutated from existing genes, the traditional breeder, farmers and society will usually be out of luck.

Today's modern biotechnology using rDNA

The new biotechnology offers important improvements on traditional methods. Recombinant DNA methods enable breeders to select, transfer or modify single genes, thereby reducing the time-consuming and labor-intensive need to select out the undesirable genes and it also allows the acquisition of useful genes from any species.

Food crops

The first rDNA whole food on the American market was the *Flavr-savr*™ tomato, produced by Calgene and released in the United States in 1994. Health Canada approved use of the tomato for human food in February 1994 but it was not marketed here. Other companies also developed various GM tomatoes along with corn, soy and other crops. The Appendix (p. 51) shows the GM crops that have been approved by Health Canada and Canadian Food Inspection Agency (CFIA) to date.

Some critics argue that current GM crops benefit only the companies producing them and, perhaps, the farmers but offer nothing for consumers. While it is true most of the first GM crops carry simple, single-gene improvements like resistance to pesticides that directly benefit farmers and chemical companies, they also benefit consumers. We consumers benefit from fewer weed and insect contaminants in the food (weeds not only diminish crop yields, they also harbor insects, microbes and associated toxins) and lower prices (from enhanced production). Most consumers say they benefit from, and would support, a food product made with fewer chemicals. Farmers growing GM crops use fewer pesticides than those growing non-GM crops and the ones they do use are more environmentally benign. This also results in lower production costs, which means lower food costs in the market.

Resistance to herbicides

Many of the first GM crops display genetic resistance to a herbicide. Some critics complained that this would lead to an increase in use of chemicals on farms because the farmer would have to spray the crops with a particular herbicide although society would prefer farmers to use less, not more chemicals. However, recent analyses by the US Department of Agriculture (USDA) and others have shown a drop in the use of herbicides with the GM crops. Why? Current agricultural production often necessitates the use of several different herbicides to control the wide range of weed species present in a field. With herbicide-resistant GM crops, farmers are able to grow weed-free fields by applying a single herbicide instead of spraying several different chemicals. Farmers benefit by having to buy fewer chemicals and spraying less often (thus saving time, fuel and wear and tear on equipment) plus the crop is of higher quality, due to decreased competition from weeds. Consumers benefit from lower prices (due to enhanced production and lower input costs) and a product of higher quality (from fewer weed, insect and microbial contaminants). The environment benefits because farmers burn less tractor fuel and spray reduced quantities of chemicals—typically newer, safer chemicals that leave less residue and reduce runoff problems.

Insect resistance

A commonly used gene confers resistance to insect attacks. A gene from the bacterium *Bacillus thuringiensis*, usually called simply "Bt," produces a crystal protein toxic to certain insects but safe for humans and other animals. The insecticide Bt has been sprayed on crops to control insects for half a century. Now, by inserting the responsible gene directly into the crop plants, farmers need not spray the Bt into the environment, where it might affect innocuous insects. When the Bt crystal protein is produced only inside the plant, only those insects feeding on the plant, *i.e.* the pests, are exposed to, and killed by, the insect-specific toxin. Not only are other, innocuous insects spared, the farmer saves money, fuel and time from not having to spray.

Much of the crop in developing countries is lost to insects, even after harvest as insects quickly devour stockpiled grain. If farmers grow Bt-containing GM crops, more grain will be available to feed humans, not insects, and poorer countries will become more self-reliant.

Disease resistance

Plant diseases are usually caused by viruses, bacteria or fungi. Using GM technology, American scientists saved the papaya industry in Hawaii. Papaya ring-spot virus, which devastates the fruit and for which there is no cure, was making its way across Hawaii, destroying the papaya industry as it went. Dr Dennis Gonsalves and his team produced a "vaccine" against the virus by inserting a portion of the genetic material from the virus into papayas. The "vaccinated" papayas were immune to the virus and kept producing fruit after the virus had destroyed all the other papaya trees in the region. Indigenous farmers and the entire Hawaiian papaya industry owe their continuing livelihood to Dr. Gonsalves's team and GM technology. Similar approved products include virus-resistant squash and potatoes.

Canadian scientists are leading the way in researching biotechnology to fight fungal diseases. Dr. Santosh Misra and her colleagues at the University of Victoria (British Columbia) have developed a promising biotechnology to protect potatoes from fungal disease. Considering the problems that potato farmers in Prince Edward Island have had marketing their potatoes internationally due to the mere presence of fungal infections, successful adoption of Dr Misra's potatoes would be a great boon to the potato producers. They should also be welcomed by Prince Edward Island's consumers and environmentalists because the protected potatoes would require less treatment with chemical fungicide, resulting in less runoff of chemicals into the environment and less exposure for humans.

Delayed ripening

Other products include the delayed-ripening *Flavr-Savr*™ tomato, which was modified not by inserting a foreign gene but by altering a naturally occurring tomato gene. Polygalacturonase (PG) is an enzyme controlling the ripening of tomatoes (and other fruits). By reinserting a normal tomato PG gene into the tomato in reverse orientation, Calgene scientists found they could delay the expression of PG in the tomato, slowing the ripening process. The delayed ripening meant the GM tomatoes had a prolonged shelf life and also facilitated harvesting and transport to markets.

Enhanced quality

The first nutritionally enhanced GM foods are on the market and many more are on the way. DuPont made a soybean with a modified oil composition, increasing the desirable monounsaturate oleic acid to over 80% of the oil from about 24%. Several companies and public institutions are developing soy, canola and other vegetable oils with reduced saturated fat content and other enhancements.

Other uses of molecular genetic technologies do not depend on gene transfer or modification. For example, early detection and treatment of crop disease is an important strategy for reducing chemical inputs in farming. Biotechnology offers highly sensitive and selective diagnostic tools to detect low levels of infection or infestation. This is essential for speeding the delivery of control measures before the problem grows. Say a fungal disease, for example, begins to infect a crop field. Using sensitive assays, the farmer can identify and treat only the infected portion of the field long before any real damage is done, ensuring the highest quality of the crop and precluding the need for chemical treatment (or loss) of the entire crop. Field kits based on antibodies that specifically recognize a disease or pest agent are already in use for soybean root rot and certain bacterial diseases of tomatoes and grapes. DNA fingerprinting—famous for its use in forensics—is also used in identifying plant diseases and infestations.

Processed foods
Processed foods have gained acceptance among time-conscious consumers in search of easily prepared, nutritious and flavourful meals. Three components of the food-processing system are targeted by new biotechnologies: (1) nutritional and chemical composition (such as proteins, fats and carbohydrates); (2) bacteria and yeast used in fermentation and other processes; and (3) enzymes used to enhance colour, flavour and texture. Starter cultures of bacteria and yeast are the mainstay of much of the processed-food industry. Biotechnology has contributed substantially to microbial genetics, improving our understanding of bacterial and yeast genes that are involved in making foods as diverse as bread, yogurt, cheese, wine and beer. Using modern genetics, food processors are culturing new genetically modified strains of microorganisms that have combinations of enzymes useful in food processing.

Biotechnology is also useful in the isolation and production of enzymes used directly in food processing. For example, the enzyme amylase affects the texture and freshness of bread dough. Chymosin, the active enzyme in rennin extracts (which are isolated from calf stomach), curdles milk to make cheese. Purified chymosin produced through genetic engineering now accounts for the most enzyme used to make cheese. It allows both vegetarians and those who follow kosher rules access to cheeses. Other enzymes, like proteases and lipases, are used to provide the aged quality of cheese. There is also an effort to help meet consumer demands for foods that can be kept fresh without synthetic additives or special packaging materials. Modern molecular methods are being used to produce substances that eliminate bacterial contamination: propionic acid to reduce fungal contamination, trehalose sugars for dried and frozen foods and antioxidant enzymes to prevent the formation of free radicals.

Tomorrow's biotechnology—anticipated products and benefits

Products for farmers

Drought, flooding, soil salinity, frost and other environmental stresses take heavy tolls on agricultural production in Canada and around the world. Several naturally occurring genes involved in plant responses to these stressors have been identified and isolated. A leading scientist in this regard is Canadian Dr Larry Gusta, at the University of Saskatchewan. Dr Gusta and his team have successfully added several genes conferring resistance to heat, cold, salt and other stresses to important Canadian crops like canola, flax and potato. These GM plants are now in field trials to ensure they display enhanced stress tolerance as they are supposed to do and carry no detrimental health or environmental effects. Similarly, Dr Eduardo Blumwald and his team at the University of Toronto developed genetically modified tomatoes able to grow in saline conditions. Much of the world's farmland is becoming saltier due to irrigation and some arid areas are naturally salty. Most crop plants suffer from excess salts but Dr Blumwald's tomatoes seem to thrive in salts that would dramatically impair ordinary tomatoes. The relevant gene is expected to confer salt tolerance when transferred to other crops also. Because these environmental stresses cause a huge loss of food products, even modest success will have enormous benefit.

Products for consumers

More consumer-oriented GM products will soon appear on our shelves. These will include nutritionally enhanced foods, such as fats and oils with a reduced saturated fat content (as mentioned above), potatoes that absorb less fat in frying, and "sugar" (fructans) with fewer calories from GM sugarbeets.

In addition to foods fortified with healthier ingredients, naturally occurring allergenic and antinutritional compounds will be reduced or eliminated. Some people, for example, are allergic to peanuts and, given the preponderance of peanuts in so many snacks, baked goods and other common foods, have to deal with the anxiety of encountering them on a daily basis. Imagine their relief if peanuts can be made non-allergenic, perhaps by removing one or more genes for allergenic proteins. Similarly, many foods naturally produce antinutritional compounds, such as alkaloids or cyanogenic glycosides. Antinutritional substances can be toxic or may interfere with normal nutrition. Although modern breeding has reduced the naturally occurring hazardous chemicals in our food crops, they have not been eliminated completely. While current procedures monitor levels of these undesirable compounds in new crop varieties produced by both traditional and modern techniques, GM may be used to eliminate or reduce the capacity of plants to produce such compounds.

Foods with Improved quality are also coming to less fortunate peoples. Cassava is a major food for millions of people but is not especially nutritious because of the poor quality of its protein. It also contains toxic cyanogenic glycosides. Genetic modification could be used to reduce the antinutritional factors and, at the same time, enhance the quality of the protein in this food.

Over 180 million children, mostly in developing countries, suffer Vitamin-A deficiency; some 2 million die from it each year. About a billion people suffer anemia from iron deficiency. Genes producing Beta-carotene, a precursor of vitamin A, have been inserted into rice, the most important crop species in the developing world, but one deficient in this crucial nutrient. The GM rice produces beta-carotene, giving it a bronze-orange appearance, hence the name "Golden Rice." This nutritionally enhanced GM rice, generated by a team led by Dr. Ingo Potrykus in Switzerland, is being distributed, free of charge, to public rice-breeding institutions around the world. Local breeders will incorporate the new rice trait into local rice varieties for growing by local farmers. Millions of people suffering anemia from iron deficiency, blindness or other manifestations of vitamin A deficiency, will soon be able to overcome these disabling conditions at little or no additional cost.

GM food plants are also being developed to deliver vaccines. Poor people in developing countries suffer unnecessary afflictions because they lack medical treatment. Conventional vaccines are often difficult to deliver because they require expensive production, sterilization and refrigerated transport mechanisms. Tropical fruit like the banana or even temperate crops like the potato and the tomato can be genetically enhanced to produce vaccines, and they might be grown locally and more easily transported for local consumption. Current cost for delivery of a conventional vaccine injection into a patient in remote areas is about $125 per dose. Estimated cost for a vaccine delivered via bananas is about $ 0.02. Vaccinating against Hepatitis, Typhoid fever and Cholera would dramatically improve the lot of many people.

Closer to home, vaccines delivered via GM fruits might be more efficacious and also easier to deliver, especially for the very young or elderly where conventional vaccination systems are less effective.

GM animals and advances in animal husbandry

GM animals are not too far behind GM plants. We have all heard of Dolly the Scottish sheep, a clone derived from a single mature cell from her mother. Cloning is not really genetic modification as cloning implies a genetically identical copy rather than a genetically altered version but both are products of modern technology. Cloning sheep and other animals can allow us to increase populations of nutritionally enhanced or more productive animals, without going through a lengthy and expensive breeding program. The first GM (in contrast to cloned) whole animals likely to reach the market will be fish. Newfoundlander Dr Garth Fletcher's A/F Protein Canada Inc., is leading several firms around the world in producing fast-growing GM salmon in controlled fish farms. The fish are sterile so that, even if they do escape to the wild, they will not be able to reproduce.

On the production side, conventional farm animals often suffer diseases. We are all aware of the recent outbreaks of "mad-cow" disease and foot-and-mouth disease in Europe. GM vaccines can be delivered to animals via their feed, preventing disease and saving the expense of unnecessary bills for veterinarian treatment. Unfortunately, many consumers in Europe were misled into believing that biotechnology gave rise to these diseases in the first place (although there is no evidence to support the contention); hence their reluctance to accept improved technology now that would help them overcome these devastating problems.

Veterinary medicine

By identifying and mapping genes, veterinary scientists are able to understand and help correct the physiological systems that underlie animal diseases. The identification of genes involved in serious inherited diseases helps animal breeders select the healthiest animals and improve the characteristics of their herds. Genetic diagnostic kits are becoming commonplace in animal husbandry. New biotechnology vaccines are also playing important roles in veterinary medicine both in Canada and in developing countries where more stable systems for producing meat and other animal foods are of critical importance. One genetically engineered vaccine successfully controls Rinderpest, a viral disease that periodically destroys entire cattle herds in Africa and Asia. Another biotechnology vaccine has proved effective against the rabies virus in American trials. Genetically modified vaccines control everything from ticks on cattle to *E. coli* infections in piglets and calves. VIDO, the Veterinary Infectious Disease Organization in Saskatoon, is a world leader in the development of animal vaccines using biotechnology.

Animal growth and food production

The genetic engineering of farm animals is still at an early stage of development. Most research (with a few exceptions, such as bovine and porcine somatotropin) is preliminary and will require extensive development before genetically engineered meats reach the marketplace. However, we can foresee genetic engineering eventually being used to improve the metabolic efficiency of animals, thereby enhancing the utilization of feed and improving the quality and production of meat and milk.

Advances in food processing

Other researchers are developing methods for enhancing the vitamin-C and vitamin-E content of processed foods. Strategies derived from our knowledge of the mammalian immune system are being applied to food production. The immune system attacks foreign molecules in the body by producing highly selective antibodies that identify and help destroy invaders. Exploiting that defence system, food scientists have developed antibodies that target specific food contaminants that are potentially toxic or pathogenic. These new detection systems are being tested and packaged in easily used kits. Food processors and handlers are able to identify and eliminate contaminated foods before they reach supermarket shelves.

Helping the developing world

We have discussed the advent of nutritionally enhanced Golden Rice and the utility of using GM plants to deliver medicines. Other projects will have especially important effects in poorer countries, where most of the world's population growth will occur. Food is almost always in short supply. A recent report (Sheehy, Mitchell & Hardy 2000) announced the genetic modification of rice to make it more efficient at converting sunlight to food through photosynthesis. The GM rice yields about 20% more grain than conventional rice at no extra cost to the plant, to farmers or to consumers. David Dennis, a Canadian scientist with Performance Plants Inc. of Kingston, Ontario, and Saskatoon, Saskatchewan, is also developing GM plants with improved capacity to harness sunlight.

Some crops are productive and nutritious but are deficient in specific nutrients. The varied and balanced diet we enjoy in developed countries often makes up for these deficiencies but many people do not enjoy the luxury of a diverse diet. Grain legumes, beans and peas, for example, are low in the amino acids cysteine and methionine. People forced to eat a diet predominantly of grain legumes can suffer deficiencies. Genetic modification can be used to insert genes for proteins rich in the missing nutrients to provide a more balanced diet without adding other foodstuffs.

Industrial products

Biomaterials such as bioplastics made from GM plant starch can be just as versatile as plastics produced from fossil fuels. The GM plastics are superior in that they will be made from a renewable resource and, being biodegradable, will also be more environmentally friendly.

Certain plants produce oils quite similar to diesel fuel. GM might be used to modify plants to produce fuel more efficiently. Beyond food, industrial products are also targets of genetic modification. The characteristics of plant fibers are often genetically controlled: textiles like cotton and linen (from flax) can be modified to increase quality and wearability at a lower processing cost. While Canada is a world leader in flax production, almost none is used for linen but is grown for its oil instead. Genetic modification might be applied to our flax varieties, however, to make them more suitable for textiles and other fibers or to develop varieties that produce their own pigments, thus saving materials and labour now needed to dye them.

Concerns about GM foods and products

Biodiversity—enhancing yield, sparing primitive forests

The world's population, currently six billion people, is expected to grow by an expected additional two billion over the next 20 years. Much of the growth is occurring in developing countries where local capacity for food production is seriously unstable because of poverty, political disruption, climatic stresses, soil erosion, pests and disease. The pace at which primitive forests and other natural lands are being converted to food production is increasing but this need not be the case. When biotechnology is used in combination with other strategies, it can help us address several of the central problems. The new molecular tools are both enhancing our understanding of the range and importance of biodiversity and supporting strategies that will spare and even return land to natural habitats while helping to feed the world's peoples.

The Green Revolution of the 1950s, 1960s and 1970s, with its introduction of high-yield wheat and rice varieties, proved that carefully targeted plant breeding can substantially improve local food production. The new genetic tools can both enhance and extend those improvements by delivering technological enhancements directly—in the seed. Biotechnological research is underway in agricultural centers in South and Central America, the Caribbean, Africa, Asia and the Pacific Islands. Much of this research focuses on crops that are important for indigenous farmers: rice, beans, maize, squash, melons, cassava, papaya, sorghum, potatoes and sweet potatoes. Superior varieties will help farmers achieve greater yields of nutritionally enhanced crops with lower inputs on less land.

Biotechnology offers novel products to replace those derived from forestry or agriculture. Biomaterials are under development to reduce our reliance on products based upon plant fibers and petroleum. Plant cells can be grown in large-scale vats, or bioreactors, to make products ranging from oils to flavourings and amino acids without cultivating land. These approaches all help to reduce the need to bring greater acreage under cultivation and reduce the need to harvest the world's forests.

Sustaining biodiversity

The world today supports a remarkable diversity of plant and animal life. For centuries, plant and animal breeders have sought the best traits from wild species and genetically integrated them into domesticated crops and food animals. Extensive seed banks—literally, gene banks—have been established to collect a broad range of seeds and other germ plasm both to document and characterize the world's various species and to support future breeding needs. Thus, an appreciation and utilization of biodiversity has been a hallmark of traditional plant and animal breeding programs around the world.

Biotechnology offers several strategies for sustaining and utilizing the world's biodiversity. First, it offers tools for identifying and characterizing living organisms at the genetic level. Molecular diagnostic techniques enable scientists to distinguish between, and compare species with remarkable precision. Used with traditional techniques, genetic-engineering techniques are expanding our knowledge of the range and evolution of organisms living in North American meadows and tropical rain forests alike. They allow researchers to monitor and track changes in specific populations over time. Knowledge of the genetic composition of wild species also enables breeders to identify and make use of genes that encode traits that are beneficial for food production.

Some critics argue that by promoting monoculture (i.e., intensive growth of one species), GM contributes to the problem of reduced biodiversity in the world. The opposite is true. Adding genetic information does nothing to diminish what is already present. Adding a gene to a genome does not delete any genes from the species, so biodiversity is maintained. Molecular genetic technology is being applied to identify and characterize genes in many species, thus helping to establish seed banks and gene banks to ensure the preservation of biodiversity. The genome of rice was recently analyzed and the information is being made available to breeders around the world. This is an important advance not just for rice breeders but for all crop breeders, because many genes responsible for food production and quality are present in other food crops as well as rice. The genomes of other major crops are also being analyzed and this additional knowledge will facilitate even greater contributions to food production and nutrition. Canadian scientists are actively contributing to the molecular gene analyses (called genomics) for several of our most important crops, including wheat, barley and canola.

Biodiversity is not at risk from genetic modification but from natural environments being converted to farmland to provide more food for our growing populations. By growing more productive GM crops, less wild land is destroyed to make farmland. Parks and refuge areas can be left undisturbed for us all to enjoy and for biodiversity to flourish.

The public debate

The public debate over the benefits and hazards of genetic technology suffers from an astounding array of misinformation, misunderstanding and manipulation. Unsubstantiated scare stories abound.

View of scientists

Most scientists knowledgeable about genetic engineering support this technology. They know much of the negative information in the public debate is based on false assumptions, and they know that the tremendous potential benefits far outweigh the manageable hazards. But, most importantly, they understand that risks are associated with the products, not with the methods by which the products are made. That is, genetic modification is only a process to make a certain product, just like cooking is a process used to generate healthy meals. In each case, the process might be used to make hazardous products instead.

Dr. C.S. Prakash from Tuskegee University was so upset by the poor state of the public debate on GM foods that he and his colleagues drafted and circulated a petition. It began:

> We, the undersigned members of the scientific community, believe that recombinant DNA techniques constitute powerful and safe means for the modification of organisms and can contribute substantially in enhancing quality of life by improving agriculture, health care, and the environment. (www.agbioworld.org/declaration/petition/petition.phtml)

The petition, entitled *Declaration of Scientists in Support of Agricultural Biotechnology*, has been signed by over 3000 scientists worldwide, including Nobel laureates James Watson, who jointly discovered the structure of DNA in 1953, and Norman Borlaug, the 1970 winner of the Nobel peace prize and father of the "Green Revolution." At the same time, scientists are aware of the potential for undesirable features in certain GMOs and carefully assess all new GM products for any sign of unexpected or unintended results.

Scientists and regulators are aware that there are real hazards associated with all technology, including biotechnology and traditional technologies. In addition to the considerable efforts to identify and manage the real risks of biotechnology, the scientific community also understands that many citizens have social and ethical concerns about some aspects of biotechnology. These concerns too are being respected and addressed. Several Canadian universities and granting agencies are including sociologists, ethicists and other non-technical experts on faculty and staff to add their perspectives to technological developments and projects that at one time would be the exclusive domain of the scientists.

In addition, scientists themselves are helping quell anxieties about technology. For example, there is considerable disquiet about research into human stem cells. Although almost everyone agrees that stem cells hold tremendous potential for various medical treatments, many are concerned with the ethics of sacrificing human embryos to obtain the powerful cells. At the Montreal Neurological Institute, Dr Freda Miller and her team has discovered a method of collecting stem cells from skin, thus avoiding the need to destroy embryos. Scientists can now continue their experiments to develop new treatments using the rare and precious stem cells, comforted in knowing no human embryos were destroyed to provide them.

The safety of GM foods

GM products are not inherently hazardous. We have been using GM to make pharmaceuticals for a quarter century, with no documented cases of harm attributable to the genetic-modification process. Three hundred million North American consumers have been eating several dozen GM foods grown on over 100 million acres since 1994. Again, there are no documented cases of harm attributable to the process by which the GM crops were bred. In early 2000, the Organisation for Economic Cooperation and Development (OECD) invited 400 world experts, including academic researchers and representatives of government and industry as well as environmental activist groups to a conference on the safety of GM foods. Groups adamantly opposed to GM foods were given the chance to present evidence to support their assertions. They were unable to cite any cases of harm from GM foods.

In April 2000, the US House of Representatives Committee on Science, subcommittee on Basic Research, released the report, *Seeds of opportunity: an assessment of the benefits, safety and oversight of plant genomics and agricultural biotechnology*, which concluded that there is no significant difference between plant varieties created using agricultural biotechnology and similar plants created using traditional crossbreeding. In the same month, the US National Academy of Sciences released a report of their blue-ribbon group, the Committee on Genetically Modified Pest-Protected Plants, to study the matter and issued their own statement: "The committee is not aware of any evidence that foods on the market are unsafe to eat as a result of genetic modification" (National Academy of Sciences Committee on Genetically Modified Pest-Protected Plants, National Research Council 2000). GM foods have been consumed by many humans for six or seven years and any inherent problems with genetic modification as a technology would have been revealed during this time. But, not one problem has been documented.

Overall, many thousands of GMOs have been generated and tested in laboratories and field trials around the world since the mid-1970s. These include many plant, animal and microbial species modified with a range of different genes from diverse sources. Only a small proportion of these was intended for commercial release. Instead, most were developed to test the environmental and health safety of the process and the products. In spite of considerable effort to find evidence of harm from the genetic modification process, none was found. Certainly some products were identified as potentially hazardous: for example the allergenic Brazil-nut storage protein gene in GM soybean was clearly a health hazard to those allergic to Brazil nuts (see more on this below). But, in each case, the hazard is due to the nature of the specific new trait, not to the process by which it was made.

Current regulatory practice and due diligence on the part of developers identify and eliminate such products long before they get to the market. Similarly, hazardous products are occasionally created using conventional plant-breeding methods. For example, some conventional tomatoes produce too much tomatine, a potentially hazardous, naturally occurring, alkaloid. Regardless of the specific method of breeding, new foods are vigorously tested to minimize risk. Plant breeders spend eight to 12 years or more analyzing and evaluating prospective new varieties. In addition to the usual measures of seed yield, maturity, response to disease infection and other traits of interest to farmers, they also conduct chemical analyses to ensure that quality characteristics are preserved. In addition, the new lines are scrutinized for the presence of naturally occurring but undesirable compounds, like cyanogenic glycosides. Prospective new foods, whether produced by conventional or modern biotechnology, are eliminated if they show any sign of unmanageable potential hazard.

Issues in the popular press
Rats!
A scientist in Scotland, Dr. Arpad Pusztai, was testing a strain of GM potatoes by feeding them to rats. After 10 days of eating nothing but the GM potatoes, the six rats were "sacrificed" and a number of measurements taken of their

internal organs. Astonished at the preliminary results, Dr. Pusztai decided he had to share his findings with the world and went on national television before seeking peer review and before repeating the experiments. He claimed the rats fed GM potatoes suffered suppression of their immune systems and damage to their internal organs. The conclusion reached by millions of viewers was that the potatoes became toxic as a result of the genetic modification. The resulting furour spurred the prestigious Royal Society to convene an expert panel to assess the results. After considerable deliberations and review of all available data, they reported that the evidence did not support Dr. Pusztai's conclusions. They criticized almost every facet of the experiment, from design to execution and interpretation. An important point missed by most journalists was that the genetic modification introduced a gene to make lectins, a toxin. The public, relying upon the news media for information, were not told the rats were being forced to eat a toxic lectin, and that this toxin, not the method used to introduce it, might have been responsible for their illness.

Allergens

Do genes and their associated proteins change character when inserted into a different species? Although many thousands of GMOs have been produced and tested, no unexpected changes in character have been reported. One case illustrates how GM does not change fundamental characteristics. Grain legumes (beans, peas, lentils, etc.) are low in the essential amino acids, methionine and cysteine. In an effort to balance grain legumes by providing a gene for a protein rich in methionine and cysteine, scientists at Pioneer HiBred identified a gene for such a protein in the Brazil nut and transferred it to soybean. When the scientists were checking into the potential allergenicity, they found the Brazil nut was highly allergenic to some people and that the protein was responsible. When they conducted laboratory tests on the GM soybean carrying the gene from the Brazil discovered it, too, was allergenic to those people a... Brazil nut. The project was abandoned before any consumer was exposed to the soybean modified with the Brazil nut gene; it never appeared in the market and the public was not exposed to any risk. The allergenic result shows, conclusively, that the undesirable properties of the Brazil nut protein were transferred when the gene was moved into a different species. If a gene produces an allergenic protein in one species, it will likely do so in a new species. Similarly, if a gene is non-allergenic in one species, it is not likely to become allergenic when transferred to another.

While genetic modification does not change the fundamental nature of the allergenic protein, it might be able to remove allergens from food. Any parent with a child allergic to common foods knows the anxiety of trying to discern the contents of fast food, meals at friends' homes, or shared snacks at school. Imagine the relief to allergy sufferers and their parents if we can provide, using GM, non-allergenic peanuts, dairy products, cereals, seafood and other common allergenic foods. So, while GM does not increase allergies, it might be used to alleviate the problems.

Bt corn and monarch butterflies

Among the most popular traits introduced via genetic modification is insect resistance. The common bacterium, *Bacillus thuringiensis* (Bt) has been used for half a century to combat caterpillars in crops. It is still used widely by farmers, including organic growers, as a safe, natural insecticide. The bacterium produces, from a single gene, a crystal protein lethal to caterpillars and some other insects, but harmless to other animals, including humans. That gene, or enhanced forms of it, has been transferred to several crops, including corn, soy, potato and cotton. With the plant making its own Bt, the farmer need not purchase the chemical version and need not spray the crop with the chemical.

When the insect pests start eating the crop, they ingest the crystal protein and die before causing any measurable damage to the crop. Only insects that eat parts of the Bt-enhanced plants will be affected.

Monarch butterflies, as lepidopteran insects, are susceptible to the Bt toxin so when, in a preliminary laboratory study in 1999, the Monarch larvae were forced to eat pollen containing the Bt protein, they failed to thrive and some even died. This raised concern in some quarters because it was widely and incorrectly interpreted to mean that GM crops were threatening non-pest insects, like the Monarch butterfly. Several follow-up studies showed the effect of GM pollen on non-target insects, including the Monarch butterfly, to be negligible under "real-life" field situations. The scientific community discounted the original report because it was conducted under artificial laboratory conditions, the larvae were allowed to eat only corn pollen (which they do not often encounter in the open environment) and because there was no comparison group of larvae fed on ordinary corn pollen sprayed with regular Bt insecticide.

When proper experimental studies were subsequently conducted, including that of Dr Mark Sears from the University of Guelph, the concerns for the threat to the Monarch butterfly were found to be unsubstantiated.

Plants with selectable marker genes

Researchers often introduce a gene that encodes an easily detected substance that can be used as a signal or "marker" to help determine which plant cells and tissues have successfully taken up new genetic material. These selectable marker genes give a growth or survival benefit to successfully engineered cells. They are very useful because the efficiency of introducing genes into cells is relatively low. Marker genes provide researchers with a simple mechanism to separate cells that contain new genes from those that do not. Marker genes are also routinely used in combination with other genes that encode specific traits of interest. If a plant cell has taken up one gene, it is generally the case that it has taken up both the marker and the gene of commercial interest. The presence of a marker gene provides an easy indicator that the other desired gene is also present. Recently, however, some critics have called the safety of these marker genes into question.

Are marker genes safe?
Several groups are concerned about the safety of selectable marker genes in foods. The mere presence of a marker gene (or any other "foreign" DNA) is not a food-safety concern because scientific evidence shows DNA itself is not hazardous. The pertinent issue is whether the gene product, the protein made from the gene recipe, is safe. As consumers, we routinely ingest vast amounts of foreign, uncharacterized and largely extraneous genetic material (and their encoded proteins). This is a consequence of conventional plant reproduction—both natural and induced by human intervention—in all the fruits, nuts, vegetables and grains we eat every day. We have learned from this experience that new combinations of genes or entirely new kinds of genes or proteins in our foods are not, in and of themselves, indicators of risk.

Consider, as a case in point, the gene resulting in resistance to kanamycin (an antibiotic) used as a selectable marker in the production of Calgene's *FlavrSavr*™ tomato). That same gene is found in harmless bacteria that are normally present on fresh fruits and vegetables, things we eat every day without undue harm. The kanamycin resistance gene is also found in bacteria that populate the human gastrointestinal tract. In considering safety, it is no more or less relevant to know that the gene is used as a selectable marker in genetically engineered plants than it is to know that it is a normal component of bacteria that live on many of our foods or in our intestines. Safety is not determined by how or why the gene (and its product) was introduced into food; safety is determined solely by the characteristics of the gene product and our experience with it in the food supply. However, most scientists are now developing GM products without using genes resulting in resistance to antibiotics as markers.

The safety of foods developed with biotechnology

Are genetic engineering techniques in any way inherently dangerous or unpredictable? Numerous international scientific organizations, including the US National Academy of Sciences, the US National Research Council and the Royal Society (UK) have all emphasized that the new single-gene techniques are both precise and reliable. These organizations recommend that safety determinations focus on the nature of the trait that is introduced into a plant or animal. New foods developed with biotechnology are compared with similar foods produced using more traditional methods. The genetic engineering of crops and food has been more carefully scrutinized by government and university scientists than any crop-breeding technology in the history of agriculture. Over the past 25 years, millions of laboratory experiments have been conducted with rDNA techniques and with genetically modified organisms. There have been thousands of field experiments with rDNA-modified plants throughout the world. The genetic and phenotypic characteristics of every new genetically engineered plant are evaluated at each stage of development—laboratory, greenhouse and small-scale field trial—under various national and international guidelines and regulations. There is no evidence that rDNA techniques or genetically modified organisms pose any unique or unforeseen hazards to the environment or to human health. Common sense, combined with empirical evidence and observation, dictates that, compared to traditional breeding processes that involve hundreds of thousands of genes, transferring single genes greatly enhances our ability to judge risk and safety. Support for this assertion comes from an unlikely source. Recently, the European Community issued a book summarizing the results of 81 research projects, conducted over 15 years, into the safety of GMOs (Kessler & Economidis 2000). These projects, employing some 400 research teams, almost all academic and government employees, studied the safety of GM microbes, plants and animals for the environment and human health. Not one of these now completed projects provided any evidence that GMOs were any more hazardous than their conventional counterparts. Even though Europe has been the steadfast critic of biotechnology, their own scientists conducting their own research programs were unable to find any scientific justification to maintain a moratorium on products of biotechnology. Greater certainty about the genetic modification means greater accuracy in safety assessments.

The regulation and approval process

Contrary to assertions from some critics, products of biotechnology are stringently regulated before they are placed on the market. In Canada, several government agencies bear responsibility for assessing and judging the environmental and health safety of new biotech products. The Canadian Food Inspection Agency (CFIA), Health Canada, Environment Canada and the Department of Fisheries and Oceans, operating using several statutes and regulations, ensure the safety of all products of biotechnology. Most biotech products are regulated by at least two of the agencies, depending on the exact nature of the improvement. CFIA has primary responsibility for assessing the environmental effects of releasing plants with novel traits, and Health Canada is primarily responsible for ensuring food safety. CFIA also assesses any new animal feed intended for the market, even those products where feeding animals is a minor use or a by-product. The Department of Fisheries and Oceans is concerned with novel biotech products like the fast-growing salmon currently under development. And, just to ensure that nothing slips through, the Canadian Environmental Protection Act (CEPA), administered by Environment Canada, is invoked whenever a product appears to circumvent other federal legislation.

Most products require separate approvals for environmental release, for use as animal feed, for variety registration and for use as human food, involving at least two agencies, CFIA and Health Canada. The Appendix (p. 51), based on the official CFIA chart (updated regularly on their web site www.inspection.gc.ca/english/plaveg/pbo/pntvcne.shtml), shows the status of the various approvals. To date, about four dozen genetically modified plants with novel traits have been through the approval process in Canada. These include about a dozen varieties each of canola and corn, plus several potatoes, squash, soybeans, tomatoes, wheat, flax and even sugarbeet and cottonseed. Several of these will not be grown in Canada (even genetically modified cotton cannot survive here!), but needed government approval to be imported as food or feed ingredients.

The role of CFIA

The Canadian Food Inspection Agency (CFIA) is usually the first government agency to receive notice of an impending new product of biotechnology because, long before a biotechnology product is assessed for release to farmers, CFIA regulates early-stage evaluations in small confined trials. It can take 10 or 12 years of evaluations, all of which have to be successful before development can advance to the next step, to get a plant with a novel trait ready for market. Usually, the first step in developing a new plant with a novel trait is either importing it (which is regulated by CFIA) or using various breeding methods in the laboratory and confined greenhouse facilities. CFIA is aware of these new plants from the first small field trial because they are stringently regulated. Depending on the species and the nature of the modification, the trials must be grown in isolation from other plants and the likelihood of spread by pollen or seed is minimized by requiring detasseling, bags over flowers, or other means of controlling the flow of genes. The trial sites are subject to inspection by CFIA officials, restrictions are placed on how the land might be used into the future and the area must be monitored for signs of gene escape. As the years go by, if the novel plants show no signs of environmental or agronomic risk and pass the other performance tests, the size of the trials are gradually increased and restrictions reduced, until the developer requests approval for unconfined release into the environment. This is when CFIA makes a major assessment of the novel plant and its potential for harm to the environment. On a case-by–case basis, CFIA considers the biology of the plant, the nature of the modification and the likelihood that the novel plant will become more weedy or invasive or that it might adversely affect other species or that the novel genes are likely to flow to other plants and, if so, with what consequences. Only those products judged by CFIA to be as safe as their more traditional counterparts are ultimately approved.

All plants with novel traits intended to be grown in Canada must pass this rigorous scrutiny. In addition, most field crops (whether products of biotechnology or of traditional breeding) need to be approved as registered varieties, a separate and equally rigorous procedure. To have a variety registered, the developer provides a seed sample to a government agent, who arranges and oversees field trials consisting of breeding lines of the same crop type from different breeders. The developing breeder has no control over these government-sponsored trials, which test the performance of the breeding lines against standard commercial varieties grown by farmers. Only the best lines make the grade and are supported by an independent expert committee for registration as a variety.

A few products, such as cottonseed, need not undergo variety registration or unconfined release because they are grown elsewhere (usually the United States) and imported here as ingredients. This does not mean they escape regulatory scrutiny. They must, like all novel products intended for food or feed, still pass the assessments conducted by CFIA (for use as feed) and Health Canada (for use as food) before they are allowed on the market.

How does Health Canada ensure safety?

Health Canada requires that the following questions be addressed before a new food—genetically engineered or produced through traditional methods—is introduced to the marketplace:

- Does the food contain genes from known allergenic sources?

- Does it contain genes from toxigenic sources?

- Are the concentrations of natural toxigenic substances increased?

- Is the nutrient, including fat or cholesterol content, changed?

- Does the food contain a substance that is new to the food supply?

Any food that contained increased levels of a natural toxicant require approval and could be banned from the marketplace. Health Canada is also concerned about the potential for introducing an allergen into a food in which a consumer would not expect it. Foods derived from known allergenic sources must be demonstrated not to be allergenic or must be labeled to identify the source. In cases of potentially serious risk of allergenicity, the foods would be banned from the food supply.

Labeling of GM foods

Many consumers claim they want GM foods to carry a label. But, labels are intended to convey meaningful information on the contents—either for nutritional or health-related issues. Current regulations require GM foods to carry a label if the food is substantially different from similar conventional foods: if, for example, there is significantly more or less of a vitamin or if there is an allergen or otherwise toxic substance in the food. If the food is identical to regular version of the same food, however, a label would be misleading.

Some say that we must have labels to allow "consumer choice," accommodating those who, for whatever reason, do not want to buy GM foods. To these people, all foods produced or derived from a GMO must carry a label. The rationale for such a proposal, however, is important if we are to have effective labeling, even when it is simply to facilitate choice. Some consumers justify wanting labels on GM foods because they believe insertion of DNA—a chemical—into food is an adulteration or contamination of food with a foreign chemical. However, all living things—including food plants, animals and microbes—contain DNA. Their concern usually dissipates when informed that DNA is a normal component of every healthy diet; DNA is an ingredient in virtually all foods.

Many of the same people demanding a label on all GM-derived foods would support a mandatory label on any food containing "chemicals." Since all foods are composed of chemicals, a generic label "this product contains chemicals" would be meaningless, as it does not provide any nutritional or health-related information with which to make an "informed choice." Also, as it would necessarily appear on every package, it would also very quickly be ignored, nullifying the entire utility of labeling.

Well, what about using the criterion "foreign DNA" to identify those products requiring a GM label? Then, we have complications in defining "foreign," as many genes are common across many organisms. Also, a tomato engineered with a gene from a different variety of tomato would not carry any "foreign" DNA yet advocates of labeling would insist it be so labeled because it was produced using GM techniques.

A soybean carrying a bacterial gene would clearly be a GM plant with "foreign DNA" and so subject to the mandatory label. But, what about the oil squeezed from the soybean? The oil carries no DNA, "foreign" or otherwise, so, if we used "foreign DNA" as a criterion, the oil would not need to be labeled. Yet, advocates insist it would.

What if we later discover that soybean naturally carries a gene identical to the one inserted from bacteria. Many genes from many species are quite similar to each other, so this is not an outlandish possibility. Although the inserted gene came from a bacterium, it would not be considered "foreign" if the soy already carried a natural version, would it? In this case, if "foreign gene" were the criterion for labeling, the soybean, clearly GM with a gene from bacteria, would need no label.

We should also consider the case of a tomato gene inserted into a bean plant. The bean plant would be GM with tomato DNA present inside it. Advocates of mandatory labels expect such a bean to be labeled. What if we use this bean to make a casserole, complete with ordinary tomatoes? The casserole would consist of beans and bean genes along with tomatoes and tomato genes. Would the casserole have to be labeled? It is indistinguishable in every way from a casserole made with ordinary beans and tomatoes.

Labels are useful when they tell us about meaningful differences, as when a food has added calcium or contains peanuts. We use this information to make considered choices to purchase a beneficial product, like one with enhanced vitamin content, or to avoid potentially hazardous products, as consumers with peanut allergies avoid foods containing peanuts. When there is no discernible difference between two foods, labeling one of them only causes confusion and jeopardizes the value of meaningful labels.

All health authorities agree DNA by itself is a non-hazardous nutrient so a label only becomes meaningful when associated with what federal regulators call "material information," such as a specific hazard. In those cases, current regulations already assure appropriate labels.

What about the sources of genes in foods?

Some people wonder whether the source of a gene affects the safety of foods. Given the tremendous overlap in genes among humans, animals, plants and even microorganisms, and given the fundamental chemical relatedness of DNA in all organisms, the source of the gene is of limited importance to judgments on safety. Rather, information on the gene product—the function of the protein that the gene encodes, its effect on the food and the way in which that food is intended to be used—all bear importantly on the safety of the food. For example, a substantial number of people are allergic to peanuts. If a gene from the peanut is transferred into a tomato, one might reasonably worry about the potential allergenicity of the tomato. However, if the protein encoded by that gene is known not to contribute to the allergenicity of peanuts, then the new tomato will not be a problem for people allergic to peanuts. Information about the source of the gene alone is thus of minimal usefulness.

Conclusion

Modern biotechnology is being used in agriculture and food production to provide more, better and safer products. The extent to which it will be fully utilized for the benefit of consumers depends on support for innovation and improvement in farming and food production, on the one hand, and on support for scientifically sound regulatory policies that protect against tangible food safety risks, on the other. This is a delicate balance. Biotechnology using similar genetic techniques in the field of medicine and human health is well accepted by the public and professional communities as a safe and effective means to provide more and better treatments. Because agricultural biotechnology is younger and some critics remain wary, new food products will appear gradually in the marketplace over the next few years. However, with the continuing accumulation of evidence of safety and efficiency, and the complete absence of any evidence of harm to the public or the environment, more and more consumers are becoming as comfortable with agricultural biotechnology as they are with medical biotechnology. With the research pipeline filled, consumers and farmers can expect a wide variety of new products in the not too distant future.

Appendix

The information in this Appendix is derived from the tables published by the Canadian Food Inspection Agency on its web-page, Status of Regulated Plants with Novel Traits (PNTs) in Canada (www.inspection.gc.ca/english/plaveg/ pbo/pntvcne.shtml). Please see this web-page for details, notes on interpretation and updated information.

CFIA's tables include both genetically modified varieties developed using recombinant DNA techniques and those developed using other methods (e.g. Pioneer's imidazolinone-tolerant canola and Cyanamid's wheat). Most countries define GM according to the method used to develop a plant and, thus, exclude the latter. Canada's criterion, however, is based on novelty of trait and so includes new varieties expressing novel characteristics whether produced by conventional breeding, mutagenesis or recombinant DNA techniques. (See note i on www.inspection.gc.ca/english/ plaveg/pbo/pntvcne.shtml.)

The list below shows:
    product (and designation),
    novel trait(s)
    applicant at time of application

Canola

(1) CanAlan McHughen [alanmc@citrus.ucr.edu]ola (HCN92)
Glufosinate ammonium tolerance
AgrEvo Canada

(2) Canola (GT73, also known as RT73)
Glyphosate tolerance
Monsanto Canada

(3) Canola (NS738, NS 1471, NS 1473)
Imidazolinone tolerance
Pioneer Hi-Bred Int'l

(4) Canola (MS1, RF1, MS1xRF1, MS1, RF2, MS1xRF2)
Male sterility / Fertility restoration / Glufosinate ammonium tolerance
Plant Genetic Systems

(5) Canola (23-198, 23-18-17)
Higher quantities of laurate and myristate
Calgene

(6) Canola (GT200, also known as RT200)
Glyphosate tolerance
Monsanto Canada

(7) Canola (HCN28, also known as T45)
Glufosinate ammonium tolerance
AgrEvo Canada

(8) Canola (45A37, 46A40)
High oleic / Low linolenic acid
Pioneer Hi-Bred Int'l.

(9) Canola (MS8, RF3, MS8xRF3)
Male sterility / Fertility restoration / Glufosinate ammonium tolerance
Plant Genetic Systems

(10) Canola (Oxy-235, also known as Westar Oxy-235)
Oxynil (Bromoxynil and Ioxynil) tolerance
Rhone-Poulenc

(11) Canola (B. rapa) (ZSR500, ZSR502, ZSR503)
Glyphosate tolerance
Monsanto Canada

(12) Canola (B. rapa) (HCR-1)
Glufosinate ammonium tolerance
AgrEvo Canada

Corn

(1) Corn (Event 176)
European Corn Borer resistance / Glufosinate ammonium tolerance (selection system)
CIBA Seeds/ Mycogen

(2) Corn (3751IR, also known as 3417IR)
Imidazolinone tolerance
Pioneer Hi-Bred Int'l.

(3) Corn (EXP1910IT)
Imidazolinone tolerance
ICI/Zeneca Seeds

(4) Corn (X4334CBR, X4734CBR, also known as Event Bt11)
European Corn Borer resistance/ Glufosinate ammonium tolerance (selection system)
Northrup King Co.

(5) Corn (DK404SR)
Sethoxydim tolerance
BASF

(6) Corn (Liberty Link™ lines: T14, T25)
Glufosinate ammonium tolerance
AgrEvo Canada

(7)   Corn (DLL25)
      Glufosinate ammonium tolerance
      Dekalb Genetics Corporation

(8)   Corn (MS3)
      Male sterility/ Glufosinate ammonium tolerance
      Plant Genetics Systems

(9)   Corn (MON809)
      European Corn Borer resistance/ Glyphosate tolerance
      Pioneer Hi-Bred Int'l.

(10) Corn (MON810)
      European Corn Borer resistance
      Monsanto Canada

(11) Corn (DBT418)
      European Corn Borer resistance/ Glufosinate ammonium tolerance
      Dekalb Genetics Corporation

(12) Corn (MON802)
      European Corn Borer resistance/ Glyphosate tolerance
      Monsanto Canada

(13) Corn (MON832)
      Glyphosate tolerance
      Monsanto Canada

(14) Corn (GA21)
      Glyphosate tolerance
      Monsanto Canada

(15) Corn (Cornline 603)
      Glyphosate tolerance
      Monsanto Canada

Cottonseed

(1)   Cottonseed (Bollgard™ lines: 531, 757, 1076)
      Lepidopteran resistance
      Monsanto CanadaNot grown in Canada

(2)   Cottonseed (Roundup Ready™ lines: 1445, 1698)
      Glyphosate tolerance
      Monsanto Canada

(3)   Cottonseed (BXN™ lines: 10215, 10222, 10224)
      Bromoxynil tolerance
      Calgene

(4)   Cottonseed (31807, 31808, BXN-Bollgard)
      Bromoxynil tolerance / Lepidopteran resistance
      Monsanto Canada

Flax

(1)   Flax (FP967, also know as CDC Triffid)
      Sulfonylurea tolerance
      University of Saskatchewan

Potato

(1)   Potato (New Leaf™ Russet Burbank lines:
      BT06, BT10, BT12, BT16, BT17, BT18, BT23)
      Colorado Potato Beetle resistance
      Monsanto

(2)   Potato (New Leaf™ Atlantic lines: ATBT04-6,
      ATBT04-27, ATBT04-30, ATBT04-31, ATBT04-36)
      (Also includes, New Leaf™ Superior lines:
      SPBT02-5, SPBT02-7)
      Colorado Potato Beetle resistance
      Monsanto Canada

(3) Potato (New Leaf™ Y lines: RBMT15-101, SEMT15-02, SEMT15-15)
Colorado Potato Beetle resistance / Potato virus Y resistance
Monsanto Canada

(4) Potato (New Leaf™ Plus lines: RBMT21-350, RBMT21-129)
Colorado Potato Beetle resistance / Potato Leafroll virus resistance
Monsanto Canada

(5) Potato (New Leaf™ Plus line: RBMT22-82)
Colorado Potato Beetle resistance / Potato Leafroll virus resistance / Glyphosate tolerance (selection system)
Monsanto Canada

Soybeans

(1) Soybeans (GTS 40-3-2)
Glyphosate tolerance
Monsanto Canada

(2) Soybeans (A2704-12)
Glufosinate ammonium tolerance
AgrEvo Canada

(3) Soybeans (G94-1, G94-19 and G168)
High oleic acid
Optimum Quality Grains L.L.C.

Squash

(1) Squash (CZW3)
Virus resistance
Seminis Vegetable Inc.

(2)   Squash (ZW20)
Virus resistance
Seminis Vegetable Inc.

Sugar Beet

(1)   Sugar Beet (event T120-7)
Glufosinate tolerance
Aventis CropoScience

Tomato

(1)   Tomato (Flavr Savr™)
Delayed ripening
Calgene

(2)   Tomato (1345-4)
Delayed ripening
DNA Plant Technology

(3)   Tomato (1401F, H282F, 11013F, 7913F)
Delayed ripening
Zeneca Seeds

Wheat

(1)   Wheat (SWP 965001)
Imidazolinone tolerance
Cyanamid Crop Protection

Glossary

Acetolactate synthase; ALS   A plant and microbial enzyme responsible for biosynthesis of the essential amino acids leucine, isoleucine and valine. Several common herbicides attack this enzyme, causing the plant to starve. A modified form, inserted into flax, allows the crop to grow in soil containing excess residue of sulfonylurea chemicals, an agronomic problem in parts of the Great Plains.

Adenine methylase   The *dam* gene, derived from *Escherichia coli*, expresses a DNA adenine methylase enzyme. The enzyme, when expressed in anthers or pollen, results in the inability of the transformed plants to produce fertile male plants because it kills the developing pollen grain. The sterile male plants facilitate production of hybrids, as they cannot self-pollinate. See also **Hybrid**.

Aminocyclopropane carboxylic acid synthase/deaminase   Enzymes involved in the fruit ripening process, they regulate biosynthesis of ethylene. Ethylene is an endogenous plant hormone known to play an important role in fruit ripening, so manipulation of these genes can result in fruit with altered ripening periods. See also **S-Adenosylmethionine hydrolase**.

Antisense DNA a process to inactivate a gene by excising the relevant gene, turning it 180 degrees and reinserting it back into the genome. *Flavr-savr*™ tomato has extended shelf life because of an antisense polygalacturonase (an enzyme involved in ripening) gene.

Barnase An enzyme that destroys nucleic acids, thus killing the cell. When the gene is inserted into plants and activated only in pollen (as is the typical use), the plant becomes male-sterile, facilitating the development of hybrids. An associated product, Barstar, may be used to restore male fertility to the plant.

Bt Cry9c, Bt CryIIIA, BtCryIA(b), etc. Insecticidal crystal proteins produced by various strains of *Bacillus thuringiensis*, commonly used in insect-protected crops and also, as a spray, by organic and conventional farmers.

DNA The chemical carrying genetic information. It is a double helix composed of a sugar-phosphate backbone with rungs of the bases Adenine, Thymine, Cytosine and Guanine. The genetic information is coded by the specific sequence of the ATC and Gs along the molecule. The DNA from a single human cell, if extracted and pulled taut, would be about six feet long.

Enzyme A protein responsible for facilitating a chemical reaction; a biological catalyst. Most genes code for proteins with enzyme activities.

EPSP synthase The enzyme 5-enolpyruvylshikimate-3-phosphate synthase (*EPSPS*), responsible for the biosynthesis of the amino acids tyrosine, phenylalanine and tryptophan. The herbicide glyphosate inhibits this enzyme, causing the plant's death by starvation for the amino acids.

Gene A unit of genetic information carried on a given portion of a cell's DNA. The specific sequence of DNA base letters (ATC and G) provides a recipe for the cell to make a particular protein. The presence or absence of the protein contributes, either individually or in combination, a specific trait to the organism. Humans carry about 80,000 genes; a typical plant about 25,000 (see figure 1).

Gene splicing The precise joining of pieces of DNA from different sources to create a novel gene construct. See **recombinant DNA**.

Genetic engineering, genetic modification In the general sense, any change, using classical or modern breeding methods, in the genetic structure of an organism to provide an improved strain. In the specific sense, the application of recombinant DNA (rDNA) or gene-splicing technology to living organisms, usually to produce a predicted and intended improvement. See **recombinant DNA**.

Genome The total complement of genetic information in an organism.

Glyphosate oxidoreductase A modified (*goxv247*) gene from *Ochrobactrum anthropi* produces glyphosate oxidoreductase, an enzyme that breaks down glyphosate, conferring resistance to the herbicide.

Gmfad2-1 The GmFad2-1 gene codes for a delta-12 desaturase enzyme involved in the synthesis of polyunsaturated fatty acids (e.g., linoleic acid) from monounsaturated fatty acids (e.g., oleic acid) in developing seeds. The soybean GmFad2-1 gene, when inserted into soybean, can cause a silencing (co-suppression) of itself and of the endogenous GmFad2-1 gene, resulting in a soybean whose oil has an oleic acid content that exceeds

80%. Conventional soybeans have an oleic acid content of about 24%.

Hybrid Many plants experience increased productivity from hybrid vigour, the mating of different parents. Corn producers have benefited from hybrid seed for years but many crops are naturally self-pollinating and so cannot readily provide hybrid seed. Several genetic technologies have been developed to interfere with pollen development, thus ensuring pollination from other plants and giving rise to hybrid seed.

Nitrilase The product of the BXN gene, derived from the soil microbe *Klebsiella pneumoniae subsp. Ozaenae*. Nitrilase degrades the herbicide bromoxynil, thus conferring protection from this herbicide.

Papaya Ring Spot Virus; PRSV Devastating disease of these tropical fruit, responsible for destroying the entire industry wherever it hits. PRSV was defeated by GM technology, which saved the industry in Hawaii.

Phosphinothricin acetyltransferase; PAT An enzyme that breaks down phosphinothricin, a herbicidal molecule in Liberty™, Basta™ and other herbicides. The gene, originally isolated from the soil microbe *Streptomyces viridochromogenes*, was modified for optimal utility in plants.

Phytase Phosphorus from plant-based feed is not readily available as a nutrient for swine and poultry. Phytase, an enzyme, converts phosphorus from plant feed into a form more available to poultry and swine.

Polygalacturonase A common fruit enzyme responsible for softening and over-ripening. The gene can be used to interfere with normal ripening of fruit (giving longer shelf-life) by insertion into a plant either in the normal orientation (co-suppression) or in reverse orientation (antisense).

Recombinant DNA; rDNA   Gene splicing; the precise joining of pieces of DNA from different sources ("recombining" genes) to create a novel gene construct. The construct is often inserted into a host organism to produce a protein, providing the host with a new trait. See **genetic engineering**.

S-Adenosylmethionine hydrolase   The *sam-k* gene is derived from *E. coli* bacteriophage T3. When inserted into plants, this gene reduces levels of S-adenosylmethionine (SAM), which is normally converted to **1-aminocyclopropane-1-carboxylic acid (ACC)** (see above), which is the first committed step in ethylene biosynthesis. Lack of SAM for conversion to ACC results in fruit with significantly reduced ethylene and so ripening is delayed.

Stem cell   A rare type of cell found in animals (including humans) capable of growing into one of many different mature types of cell (hence its value to medical research). Most commonly found in embryos but also present elsewhere, including mature skin tissue.

Thioesterase   The thioesterase gene isolated from the California bay (*Umbellularia californica*) is the recipe for the 12:0 thioesterase enzyme. This enzyme results in the seed accumulation of the 12-carbon saturated fatty acid, laurate, used in soaps and detergents.

Transposon   A piece of DNA with the ability, under certain conditions, to remove itself from one location (*locus*) in a chromosome and re-insert itself at another. Also called a "jumping gene."

Virus coat protein Crucial protein for viruses; the gene can be used to confer immunity to the pathogen when transferred to a host organism.

Virus replicase An enzyme responsible for virus reproduction. The isolated gene, when inserted into a plant, can confer immunity to an attack by the pathogen.

References and further reading

Arntzen, C. 1997. High-tech herbal medicine: Plant-based vaccines. *Nature Biotechnology* 15: 221–222.

Arntzen, C.J. 1997. Edible vaccines. *Public-health-rep* (Boston, MA: US Public Health Service) v. 112 (3): 190–197.

Borlaug, N.E., & C. Dowswell 2000. *Agriculture in the 21st Century: Vision for Research and Development.* Digital document: http://www.agbioworld.org/articles/21century.html.

Canola Council of Canada 2001. *An Agronomic and Economic Assessment of GMO Canola. Impact of Transgenic Canola on Growers, Industry and Environment.* Digital document http://www.canola-council.org/production/gmo_main.html.

Couglan, A. 2000. Filling the bowl: For billions worldwide, a modified grain could end the lean times. *New Scientist* (April 1). Digital document http://www.newscientist.co.uk/news/news_223230.html

Gonsalves, D. 1998. Control of papaya ringspot virus in papaya: a case study. *Annual reviews phytopathol.* (Palo Alto, CA: Annual Reviews, Inc.) v. 36: 415–437.

Grace, Eric S. 1997. *Biotechnology Unzipped: Promises and Realities.* Washington DC: Joseph Henry Press.

Kessler, C., & I. Economidis (Eds.) 2001. EC Sponsored Research on Safety of Genetically Modified Organisms: A Review of Results. EUR 19884. Luxembourg.

McHughen, Alan 2000. *Pandora's Picnic Basket: Potential and Hazards of Genetically Modified Foods.* New York: Oxford University Press.

Miller, H.I. 1997. *Policy Controversy in Biotechnology: An Insider's View.* Austin, TX: R.G. Landes.

Miller, H.I. 1999. Food label follies. *Forbes Magazine* (December): 36.

National Academy of Sciences Committee on Genetically Modified Pest-Protected Plants, National Research Council 2000. *Genetically Modified Pest-Protected Plants: Science and Regulation.* Washington, DC: National Academy Press.

Potrykus, I., R. Bilang, J. Futterer, C. Sautter, M. Schrott, & G. Spangenberg 1998. Genetic engineering of crop plants. *Agricultural Biotechnology* (New York: Marcel Dekker): 119–159.

Prakash, C.S. 2000. *Scientists in support of agricultural biotechnology.* Digital document http://www.agbioworld.org/petition.phtml

Research and Commercialization Priorities Committee on Biobased Industrial Products, National Research Council 2000. *Biobased Industrial Products*.

Royal Society 1999. *Review of Data on Possible Toxicity of GM Potatoes*. Digital document available in PDF at http://www.royalsoc.ac.uk/policy/index.html.

Sheehy, J.E. (Editor), P.L. Mitchell, & B. Hardy 2000. *Redesigning Rice Photosynthesis to Increase Yield*. Elsevier Science.

Smith, Nick 2000. *Seeds of Opportunity: An Assessment of the Benefits, Safety and Oversight of Plant Genomics and Agricultural Biotechnology*. US House of Representatives Report. Digital document available at http://www.house.gov/science/.

Weksler, M.E. 2000. GM Foods: *Opportunities to Improve Human Health*. Paper from the OECD Conference on GM foods, Edinburgh, Scotland, March 2000. Digital document available in PDF at http://www.oecd.org/subject/biotech/ed_prog_sum.htm.

Ye, X, S. Al-Babili, A. Klöti, J. Zhang, P. Lucca, P. Beyer, & I. Potrykus 2000. Engineering the provitamin A (-Carotene) biosynthetic pathway into (Carotenoid-Free) rice endosperm. *Science* 287: 303–305.

Web sites of the Canadian government relevant to biotechnology

Agriculture and Agri-Food Canada, AAFC
http://www.agr.ca/index_e.phtml

Canadian Food Inspection Agency, CFIA
http://www.inspection.gc.ca/english/ppc/biotech/conse.shtml#sa

Environment Canada
http://www.ec.gc.ca/envhome.html

Fisheries and Oceans Canada
http://www.ncr.dfo.ca/lib-bibli_e.htm

Health Canada
http://www.hc-sc.gc.ca/english/food.htm#novel

Industry Canada, Biotechnology Regulatory Assistance Virtual Office
http://bravo.ic.gc.ca/biotech/main.htm

Web sites of the American government relevant to biotechnology

FDA biotechnology information
http://vm.cfsan.fda.gov/~lrd/biotechm.html

FDA biotech final consultations
http://vm.cfsan.fda.gov/~lrd/biocon.html

USDA biotechnology information
http://www.aphis.usda.gov/bbep/bp/

USDA— status of petitions
http://www.aphis.usda.gov/biotech/petday.html

US Environmental Protection Agency (microbials)
http://www.epa.gov/opptintr/biotech/

Web sites of the biotechnology industry

AgWest Biotech (Canada)
http://www.agwest.sk.ca

Aventis
http://www2.aventis.com/cropsc/cro_main.htm

The BioIndustry Association (UK)
http://www.bioindustry.org/

Biotechnology Industry Organization (USA)
http://www.bio.org/

DuPont
http://www.dupont.com/index.html

EuropaBio (EU)
http://www.europa-bio.be/

Genentech
http://www.genentech.com/Company/timeline.html

Monsanto (USA)
http://www.monsanto.com

Novartis
http://www.seeds.novartis.com/

Zeneca
http://www.zenecaag.com/resrch/f_biotec.htm